CELEBRATING THE U.S.A. WITH POETRY

Celebrating the U.S.A. with Poetry

Walter the Educator

SKB

Silent King Books a WhichHead Imprint

Copyright © 2023 by Walter the Educator

All rights reserved. No part of this book may be reproduced in any manner whatsoever without written permission except in the case of brief quotations embodied in critical articles and reviews.

First Printing, 2023

Disclaimer
This book is a literary work; poems are not about specific persons, locations, situations, and/or circumstances unless mentioned in a historical context. This book is for entertainment and informational purposes only. The author and publisher offer this information without warranties expressed or implied. No matter the grounds, neither the author nor the publisher will be accountable for any losses, injuries, or other damages caused by the reader's use of this book. The use of this book acknowledges an understanding and acceptance of this disclaimer.

dedicated to the good citizens of the United States

CONTENTS

Dedication v

Why I Created This Book? 1

One - Brave And Free 2

Two - Watchful Eye 4

Three - Oh, America 6

Four - Celebrate Your Strength 8

Five - Hope Of The World 10

Six - Kaleidoscope Of Minds 12

Seven - The Land Of The Grand 14

Eight - United We Stand 16

Nine - Tapestry Of Dreams 18

Ten - Red, White, And Blue 20

Eleven - Pursuit Of Happiness 22

Twelve - Beacon For All Of Humankind . . . 24

Thirteen - We Rise Above	. . . .	26
Fourteen - The United States, Our Shining Star	. . . .	28
Fifteen - Opportunity Knocks	. . . .	30
Sixteen - Essence Of Liberty	. . . .	32
Seventeen - Land Of Rebirth	. . . .	34
Eighteen - Unwavering Eyes	. . . .	36
Nineteen - Hand In Hand	. . . .	38
Twenty - This Great Nation	. . . .	40
Twenty-One - We Grow Strong	. . . .	42
Twenty-Two - Stands For Equality	. . . . . .	44
Twenty-Three - Freedom's Flame	. . . .	46
Twenty-Four - Ambition And Artistry	. . . .	48
Twenty-Five - Forevermore	. . . .	50
Twenty-Six - Melting Pot Home	. . . .	52
Twenty-Seven - Joy And Cheer	. . . .	54
Twenty-Eight - Sacred Place	. . . .	56
Twenty-Nine - Open Arms	. . . .	58
Thirty - Boundless Admiration	. . . . . .	60
Thirty-One - Proudly Proclaim	. . . .	62

Thirty-Two - United, Proud, And Forever
Free. 64

Thirty-Three - Vital Seed 66

Thirty-Four - Pearl 68

Thirty-Five - The Greatness Of The United
States 70

About The Author 72

WHY I CREATED THIS BOOK?

Creating a poetry book celebrating the United States of America can serve various purposes. Firstly, it allows me to express love and admiration for the country, showcasing its diverse culture, natural beauty, and historical significance. This book can help foster a sense of patriotism and unity among readers, reminding them of what makes the U.S.A. special. Additionally, it can provide a platform for marginalized voices to share their experiences, shedding light on the challenges and triumphs of different communities within the nation. Ultimately, this poetry book celebrating the United States can inspire readers to appreciate the country's unique qualities and work towards a more inclusive and harmonious society.

ONE

BRAVE AND FREE

In a land where dreams take flight,
A nation forged in freedom's light,
United States, oh, land so grand,
With open arms, you welcome each hand.
    From sea to shining sea, so wide,
Her beauty stretches, far and wide,
Mountains majestically rise,
Valleys whisper secrets, skies mesmerize.
    From amber waves of grain so golden,
To cities bustling, never frozen,
In every corner, a different tale,
Of courage, hope, and dreams that never fail.
    In the heartland, hear the song,
Of hardworking souls, steadfast and strong,

Through fields of green and rivers blue,
America, it's because of you.
    From the Founding Fathers' vision, born,
A nation of liberty, a beacon sworn,
To stand for justice, equal rights,
To fight for freedom, day and night.
    Lady Liberty, her torch so bright,
Guiding all with a beacon's light,
From Ellis Island, a promise made,
To welcome those in search of aid.
    In the stars and stripes, we find,
A symbol of hope, unity intertwined,
A melting pot of cultures, a tapestry,
A celebration of diversity.
    America, land of the brave and free,
A place where dreams can truly be,
With gratitude, we pledge our love,
To the United States, land we're proud of.

TWO

WATCHFUL EYE

In the land where dreams take flight,
Amidst the stars and stripes so bright,
There lies a nation, strong and true,
The United States, red, white, and blue.

From sea to shining sea, it spans,
With mountains, valleys, and golden sands,
Where amber waves of grain do grow,
And bustling cities never slow.

In the heartland, souls so true,
With hands that toil, their spirits renew,
For in their labor, they find their worth,
And build the backbone of this great earth.

The vision of the Founding Fathers, wise,
A land where justice never dies,

Where equal rights are held so dear,
And freedom's song is always near.
    Lady Liberty, with torch held high,
Welcomes all with a watchful eye,
A beacon of hope in times of need,
A symbol of unity and a land that's freed.
    A tapestry of cultures, woven strong,
A melting pot where all belong,
From every corner of the Earth they came,
Seeking a life that knows no shame.
    So let us pledge our love, our might,
To this land of hope, so pure and bright,
For in the United States we find,
A place where dreams can truly bind.

THREE

OH, AMERICA

In the land of dreams, where freedom soars,
A nation of hope, where opportunity pours.
United we stand, in colors unfurled,
A tapestry of cultures, a diverse world.
    From sea to shining sea, a grand display,
Mountains tall, and prairies that sway.
Golden fields of wheat, beneath a vast sky,
Whispering winds, as time passes by.
    In the heartland, the heartbeat of the nation,
Hardworking souls, the backbone of creation.
In bustling cities, where dreams take flight,
Innovators and creators, shining so bright.
    From the grand canyon's rugged embrace,
To the concrete jungles, a bustling space.

The United States, a mosaic of dreams,
Where equality reigns, and justice gleams.
    From the shores of freedom, to the stars above,
A land of courage, bound by love.
With open arms, we welcome all,
In this land of dreams, where dreams stand tall.
    So let us celebrate this land we hold dear,
The United States, where dreams appear.
A tapestry of voices, united we sing,
In this land of hope, let freedom ring.
    For in this great nation, we proudly stand,
United as one, hand in hand.
Oh, America, our hearts shall forever soar,
In your embrace, we'll love and adore.

FOUR

CELEBRATE YOUR STRENGTH

In the land where dreams are born,
Where freedom's flame forever burns,
United States, the land we adore,
A nation built on hopes and more.

From sea to shining sea it stretches,
With landscapes vast and diverse sketches,
From golden plains to towering peaks,
In every corner, a story speaks.

In the cities, a symphony of cultures,
A tapestry woven with diverse natures,
Voices blend, harmonies entwined,
The American spirit, one of a kind.

Opportunity's gate stands wide open,
For those who dare, dreams unbroken,

A land of possibilities, where dreams take flight,
United States, a beacon shining bright.
 From the fields where hardworking hands toil,
To the factories where dreams are coiled,
The spirit of resilience, strong and true,
In every heart, it beats for you.
 Equal rights, justice for all,
A nation that stands proud and tall,
Where liberty's torch forever gleams,
United States, the land of dreams.
 Oh, America, land of the brave,
Where freedom's song will never fade,
We celebrate your strength, your might,
United States, our guiding light.

FIVE

HOPE OF THE WORLD

In the land of opportunity, where dreams take flight,
A nation stands tall, bathed in freedom's light.
From sea to shining sea, a tapestry of grace,
United we stand, in this diverse and vibrant space.

From the towering peaks of the Rockies' might,
To the sun-kissed beaches, glistening in golden light.
From the bustling streets of New York's embrace,
To the tranquil plains, where the wild horses race.

In the heartland, where amber waves stretch wide,
A symphony of cultures, side by side.
From the melting pot, our strength does rise,
A nation bound by hope, where dreams materialize.

In the shadows of history, battles fought and won,
We stand as one, under the blazing sun.
Resilient and strong, we face each trial,
With justice as our guide, we walk the extra mile.

In the land of the brave, where freedom rings,
We celebrate the beauty that diversity brings.
For every creed and color, a place to belong,
A sanctuary of rights, where we all belong.

Land of the free, home of the brave,
A nation's spirit, forever engraved.
With stars and stripes, our flag unfurled,
United we stand, the hope of the world.

SIX

KALEIDOSCOPE OF MINDS

In the land of the free, where dreams take flight,
A nation of promise, where hope shines bright.
The United States, a beacon of liberty,
A tapestry woven with diversity.

From coast to coast, in every corner we find,
A melting pot of cultures, a kaleidoscope of minds.
Different tongues and backgrounds unite,
In the pursuit of happiness, we take flight.

With open arms, we welcome all,
For in diversity, we stand tall.
From the Golden Gate to the Statue of Liberty,
Symbols of freedom for all to see.

From the bustling cities to the vast prairies wide,
Opportunities abound, where dreams collide.

With grit and determination, we forge our way,
Resilience and strength, our banner we display.

    Land of pioneers and innovators bold,
Where stories of triumph and success unfold.
In fields of science, arts, and industry,
The spirit of progress, a symphony.

    Equal rights and justice, we hold dear,
Standing together, we conquer our fears.
United we stand, hand in hand,
Bound by the love for our great land.

    Oh, America, land of the brave,
May your spirit of freedom never wane.
A land of dreams, where all can thrive,
In your embrace, we find our drive.

SEVEN

THE LAND OF THE GRAND

In the land of the brave and the home of the free,
A nation united, a tapestry we see.
From sea to shining sea, diversity blooms,
In this great land, we find our rooms.
    A melting pot of cultures, colors, and creed,
United we stand, in times of need.
From every corner of the world, we came,
Seeking freedom's embrace, a new name.
    Through trials and triumphs, we've come so far,
Bearing the stripes, and the bright stars.
From the Founding Fathers' noble design,
To the dreams that still echo, yours and mine.
    The promise of opportunity, it rings true,
A land where dreams can become anew.

With each sunrise, a chance to aspire,
To build a life, to reach higher and higher.
    From the hills of New England to the Golden Gate,
From the heartland's plains to the Lone Star State,
In this great land, where dreams can thrive,
The spirit of America will forever survive.
    So let us raise our voices, loud and clear,
To celebrate this nation we hold dear.
With unity and strength, we'll always stand,
For the United States, the land of the grand.
    A beacon of hope, a nation so true,
The United States, we salute you.

EIGHT

UNITED WE STAND

In the land where dreams take flight,
Where freedom's flame burns ever bright,
There lies a nation strong and free,
A land that's called the U.S.A.
    From sea to shining sea it spans,
A tapestry of many hands,
A melting pot of colors bold,
A nation of diverse souls.
    From mountains high to valleys low,
From fields of wheat to cities' glow,
Each corner tells a different tale,
Of triumph, struggle, and travail.
    In the heart of this great nation's core,
Lies the spirit of those who came before,

The pioneers who forged their way,
To build a better world today.
    Through hardships faced with grit and might,
From dawn's first light 'til darkest night,
We rise above the trials we face,
With resilience, hope, and boundless grace.
    Opportunity's infinite song,
Echoes through the land and carries on,
For here, within America's embrace,
Dreamers find their destined place.
    Land of the brave, home of the free,
A beacon of hope for all to see,
United we stand, hand in hand,
An undying promise to this land.
    So let us raise our voices high,
And sing this nation's lullaby,
For in our hearts, forevermore,
The spirit of America will soar.

NINE

TAPESTRY OF DREAMS

In the heart of the land, where dreams take flight,
A nation stands tall, bathed in golden light.
United we stand, in diversity we thrive,
In the land of the free, where hope is alive.
    From sea to shining sea, a tapestry unfolds,
A symphony of voices, a story yet untold.
A patchwork of cultures, colors, and creed,
We march hand in hand, in unity we succeed.
    Through trials and tribulations, we have endured,
With resilience and strength, our spirit is assured.
From valleys deep to mountains high,
We conquer the odds, reaching for the sky.
    Opportunity's embrace, a beacon of hope,
In the land of promise, where dreams elope.

From every corner, they come from afar,
Seeking a better life, guided by a star.
    Justice and equality, our guiding light,
We fight for what's just, with all our might.
In the pursuit of freedom, we stand tall,
For every voice matters, one and all.
    So let us celebrate this land we hold dear,
With grateful hearts, let our voices cheer.
For the United States, a nation so grand,
A tapestry of dreams, stitched by hand.

TEN

RED, WHITE, AND BLUE

In the land of the free, where dreams take flight,
A nation stands tall, bathed in golden light.
United we stand, a tapestry of diversity,
With open arms, we embrace every possibility.

From sea to shining sea, a majestic sight,
Where mountains kiss the sky, and rivers ignite.
The spirit of progress fuels our every stride,
As we forge ahead with hope as our guide.

In amber waves of grain, our strength lies,
A testament to the toil beneath the endless skies.
With hands that labor and hearts that dare,
We build a future where dreams are our share.

From the bustling streets of New York City,
To the humble towns that paint our tapestry,

We unite as one, a symphony of voices,
Celebrating freedom, where every heart rejoices.
    Through trials and triumphs, we stand tall,
In the face of adversity, we never fall.
For justice is our compass, guiding the way,
As we march together towards a brighter day.
    Oh, America! Land of the brave and true,
We honor your spirit, red, white, and blue.
In your embrace, dreams find their wings,
As we celebrate the land where freedom rings.

ELEVEN

PURSUIT OF HAPPINESS

In the land of dreams and endless skies,
Where freedom's flame forever flies,
Stands a nation, proud and strong,
Where hearts beat to the same song.

From sea to shining sea it stretches wide,
A tapestry of cultures, side by side,
A melting pot of hopes and dreams,
A land where opportunity gleams.

From the crimson cliffs of the Grand Canyon,
To the snowy peaks of Alaska's dominion,
From the bustling streets of New York City,
To the golden fields of the heartland's serenity.

In the face of adversity and strife,
It's the spirit of resilience that defines our life,

Through battles fought and victories won,
The American spirit is never undone.

From Ellis Island to the Golden Gate,
A beacon of hope to the tired and the late,
Where dreams take flight and fears are shed,
In the land where dreams are never dead.

Land of the free, home of the brave,
Where the pursuit of happiness we crave,
United we stand, hand in hand,
In this great and diverse American land.

So let us celebrate this land so grand,
With open hearts and a helping hand,
For together we build a brighter day,
In the land where freedom will always stay.

TWELVE

BEACON FOR ALL OF HUMANKIND

In a land where dreams are born anew,
Where freedom's flame forever grew,
A nation built on hope and might,
A tapestry of stars so bright.

From sea to shining sea it spans,
A land of dreams and brave demands,
Where mountains touch the sky above,
And rivers flow with boundless love.

In every heart, the spirit soars,
A symphony of dreams and more,
For here, in this land of the free,
We find the strength to truly be.

Through hardships faced, we rise above,
With resilience, we push and shove,

For in our hearts, the fire burns,
The will to learn, the will to earn.

    United we stand, a nation strong,
Diverse in voices, joined as one,
We celebrate our melting pot,
The cultures that this land has brought.

    From every corner, creed, and hue,
We stand together, me and you,
For in this land of liberty,
We find our strength, our destiny.

    So raise the flag, let freedom ring,
In this great land, let hope take wing,
For in the United States we find,
A beacon for all of humankind.

THIRTEEN

WE RISE ABOVE

In the land of the brave and the home of the free,
Where dreams are forged with resilience's decree,
Stands a nation proud, with strength in every stride,
United we stand, with hope as our guide.

From sea to shining sea, a land of diverse grace,
Colors of the rainbow, in every smile we trace,
For in this melting pot, a tapestry unfolds,
Where every story matters, and every voice is bold.

From the concrete jungles to the rolling plains,
Where dreams take flight, with limitless gains,
We strive for justice, equality our creed,
A beacon of light, in times of dire need.

In the heart of America, the spirit soars high,
As we reach for the stars, beyond the sky,

With hands held tight, we face each storm's might,
For together we conquer, with unwavering might.
   Through trials and tribulations, we rise above,
A united nation, built on hope and love,
With freedom's anthem, our voices ring,
A symphony of dreams, to which we all sing.
   So let us celebrate this land so grand,
Where dreams are nurtured, by freedom's hand,
For in the United States, we find our home,
A nation of heroes, where dreams are known.

FOURTEEN

THE UNITED STATES, OUR SHINING STAR

In the land of dreams and liberty,
Where hope ignites a symphony,
Stands the United States of pride,
Where dreams and freedom coincide.

From sea to shining sea, it spans,
A tapestry of diverse lands,
With towering mountains, golden plains,
And cities where ambition reigns.

In every corner, stories dwell,
Of those who dared, who fought and fell,
For in this nation's beating heart,
Resilience thrives, a work of art.

From Ellis Island's hallowed shore,
To the pioneers who dared explore,

From the struggle for civil rights,
To the stars that light up the nights.
    Justice, fairness, equality,
Are pillars of this legacy,
For in this land, we all believe,
In freedom's promise, we achieve.
    From every race, religion, creed,
We weave a tapestry indeed,
United in our hopes and dreams,
In the land of endless streams.
    So raise the flag, let voices soar,
For this great land we do adore,
The United States, our shining star,
A beacon of hope, both near and far.

FIFTEEN

OPPORTUNITY KNOCKS

In the land of stars and stripes,
Where dreams ignite like city lights,
There lies a nation, brave and true,
A tapestry of red, white, and blue.

From sea to shining sea it spans,
A beacon of hope to all who can,
A symbol of freedom, a land so grand,
Where dreams take flight, hand in hand.

In the heartland fields, amber waves,
The spirit of resilience, never wanes,
From coast to coast, mountains high,
The American spirit shall never die.

A melting pot of cultures, diverse,
A kaleidoscope of voices, converse,

In unity, we stand, shoulder to shoulder,
For liberty and justice, we grow bolder.

    From Ellis Island to the Golden Gate,
Opportunity knocks, it's never too late,
To chase your dreams, to find your fate,
In this land of promise, where dreams await.

    Oh, America, land of the free,
A shining example for all to see,
With boundless possibilities, we strive,
To make this nation truly thrive.

    So let us celebrate, with hearts aflame,
The United States, a nation's name,
With love and pride, we shall proclaim,
Long live the land of the brave and the free.

SIXTEEN

ESSENCE OF LIBERTY

In the land of the brave and the home of the free,
Where dreams are born and hearts run wild and free.
United we stand, a tapestry of diversity,
A nation built on justice, equality, and unity.

From the amber waves of grain to the shining sea,
America, a land of endless possibility.
In every corner, a story to be told,
Of resilience and courage, of legends of old.

From the bustling streets of New York City,
To the golden shores of California's vicinity.
From the snowy peaks of the Rockies so grand,
To the heartland's fields, stretching across the land.

In the melting pot of cultures, we find our strength,
A symphony of voices, harmonizing at length.

For here, in America, dreams are set alight,
Where the pursuit of happiness takes flight.
   With each passing day, we march forward as one,
An indomitable spirit that cannot be undone.
Through trials and triumphs, we've come so far,
United we stand, beneath the stripes and stars.
   So let us celebrate, this land so grand,
With grateful hearts and steadfast hands.
For in America, the land of the free,
We find the true essence of liberty.

SEVENTEEN

LAND OF REBIRTH

In the land of dreams and liberty,
Where hopes soar high and hearts are free,
Stands a nation, bold and strong,
United in diversity, where all belong.

From sea to shining sea, it spans,
A tapestry of cultures, held in hands,
Each thread distinct, yet intertwined,
Weaving a fabric, one of a kind.

In the city streets and rural plains,
A symphony of voices, different refrains,
Whispering stories of struggle and might,
Echoing the spirit that ignites.

From the towering skyscrapers that gleam,
To the rolling hills and flowing stream,
Nature's beauty, a canvas divine,
Mirroring the spirit, pure and fine.

With open arms, it welcomes all,
In this land, where dreams stand tall,
Where justice and equality prevail,
And compassion's light will never fail.

The United States, a beacon of light,
Guiding the world through the darkest night,
With courage and strength, it leads the way,
A nation united, come what may.

So let us celebrate this land so grand,
With grateful hearts and hands in hand,
For in its embrace, we find our worth,
The United States, the land of rebirth.

EIGHTEEN

UNWAVERING EYES

In the land of stars and stripes, we stand tall,
A nation united, together we enthrall.
Diverse in hues, cultures, and creed,
The American spirit, a symphony indeed.
    From sea to shining sea, our beauty unfolds,
Mountains, deserts, and plains, stories untold.
Land of opportunities, dreams take flight,
In pursuit of happiness, we find our light.
    Through trials and tribulations, we rise,
Resilient souls, with unwavering eyes.
Justice our guide, equality our aim,
In the face of adversity, we proclaim.
    A melting pot of voices, harmonious and strong,
United we stand, as we sing freedom's song.

With open arms, we welcome the world,
A tapestry of dreams, forever unfurled.
   From Ellis Island to the Golden Gate,
Hope and courage, our eternal fate.
Land of the brave, home of the free,
The United States, our sanctuary.
   Oh, America, land of dreams and might,
With gratitude, we bask in your light.
For you are the beacon, a symbol so grand,
A nation that inspires, across the land.

NINETEEN

HAND IN HAND

In the land of dreams and endless skies,
Where freedom's flame forever flies,
There lies a nation strong and true,
The United States, red, white, and blue.
    From sea to shining sea it spans,
A tapestry of diverse lands,
With mountains tall and rivers wide,
And prairies stretching far and wide.
    In cities bustling with life's embrace,
A melting pot of every race,
Where cultures blend and stories unfold,
A symphony of colors, bright and bold.
    From New York's streets, so full of life,
To California's golden shores, so rife

With dreams and hopes, that never cease,
This land of opportunity brings peace.
    Through trials faced and battles fought,
A nation's resilience never naught,
With justice sought and equality,
America stands, proud and free.
    So let us celebrate this land,
Its boundless spirit, forever grand,
United we stand, hand in hand,
In the United States, our promised land.

TWENTY

THIS GREAT NATION

In the land of dreams and vast expanse,
Where diversity and hope advance,
Lies a nation, strong and free,
A tapestry of possibility.
    From the mountains high to the plains,
Where golden wheat and cornfields reign,
The beauty of this land so grand,
A testament to nature's hand.
    In cities bustling, lights ablaze,
People from all walks of life amaze,
A melting pot of cultures, united as one,
In this land of the rising sun.
    From sea to shining sea, we stand,
A beacon of hope across the land,

The spirit of freedom forever strong,
A chorus of voices, a resounding song.
 From the struggles of the past we've learned,
Through trials and triumphs, bridges burned,
We rise with resilience, hearts entwined,
In the pursuit of justice, we find.
 Opportunity knocks on every door,
For those who dare, for those who explore,
A land where dreams can come alive,
Where passion and ambition thrive.
 In this great nation, let us unite,
In celebration of our shared delight,
For America, we proudly say,
Land of the free, home of the brave.

TWENTY-ONE

WE GROW STRONG

In the land where dreams are born,
A nation's spirit shines like morn.
United in diversity, we stand tall,
A beacon of hope for one and all.

From sea to shining sea, we roam,
A tapestry of cultures, we call home.
Hand in hand, we build our dreams,
In this land of endless streams.

Through hardships faced, we rise anew,
Resilient hearts, red, white, and blue.
With every challenge, we grow strong,
In unity, we'll right the wrong.

Opportunities boundless, skies open wide,
Here, aspirations take their stride.

From the bustling cities to the countryside,
The American dream, we'll never hide.
    In pursuit of justice, we take a stand,
Guided by freedom's steadfast hand.
Equality and liberty, our guiding light,
A nation's promise, shining bright.
    Oh, America, land of the free,
Your spirit fills our hearts with glee.
In your embrace, we find our way,
Forever grateful, we proudly say.
    So let us celebrate this land we cherish,
Where dreams are born and hopes flourish.
From coast to coast, with love and pride,
United we stand, America's stride.

TWENTY-TWO

STANDS FOR EQUALITY

In the land of liberty, where dreams take flight,
Resides a nation, shining ever bright.
United we stand, in diversity,
A nation built on opportunity.
From sea to shining sea, a vast expanse,
Where dreams are born and given a chance.
The land of plenty, where hope abounds,
Where resilience and strength know no bounds.
From the towering cities that touch the sky,
To the rural towns where fields stretch far and wide,
This land of promise, where dreams can ignite,
A beacon of hope, shining through the night.
In pursuit of justice, we strive to be,
A nation that stands for equality.

With hearts united, we face the storms,
In the pursuit of truth, our spirits are reborn.
    Oh, America, land of the free,
Where every person can chase destiny.
With open arms, we welcome all,
In this great nation, where dreams stand tall.
    From the sacrifices of the brave and true,
To the values that guide us, strong and through,
We celebrate the spirit of the free,
In this great land, the United States of America, we shall be.
    So let us raise our voices high,
And sing of a nation that will never die.
For in this land of hope and grace,
The American spirit will forever embrace.

TWENTY-THREE

FREEDOM'S FLAME

In the land of dreams and endless skies,
Where hope and freedom forever rise,
Stands a nation proud, strong, and true,
A beacon of light, the red, white, and blue.

From sea to shining sea, a tapestry unfolds,
A symphony of cultures, stories yet untold,
Where diversity thrives, in unity we stand,
In this great nation, the United States of land.

Through valleys and mountains, rivers so grand,
Nature's beauty embraced by the American hand,
From the golden deserts to the great forests wide,
This land's breathtaking splendor fills us with pride.

In the pursuit of justice, our spirits unite,
Equal rights and opportunities, shining bright,

For every American, a chance to succeed,
In this land of dreams, where dreams take the lead.
    With resilience and strength, we face every test,
For freedom's flame burns deep within our chest,
In the face of adversity, we never retreat,
United we stand, with hearts full of beat.
    Oh, America, land of the brave and free,
Your spirit, your essence, forever shall be,
A symbol of hope, a beacon of light,
A land where dreams take flight, day and night.

TWENTY-FOUR

AMBITION AND ARTISTRY

In the land of dreams and endless skies,
Where freedom's flame forever flies,
There lies a nation proud and true,
The United States, red, white, and blue.

From sea to shining sea it spans,
With mountains high and golden sands,
From rolling plains to city lights,
A tapestry of diverse delights.

In California's sun-kissed glow,
Where dreams are born and fortunes grow,
To New York's bustling city streets,
Where ambition and artistry meet.

In the heartland's fields of gold,
Where stories of resilience unfold,

To the bayous of the southern states,
Where music and culture radiates.

From the Rockies' majestic peaks,
To the Great Lakes' shores that nature seeks,
In the deserts' fiery heat and sand,
Where strength and determination expand.

United we stand, hand in hand,
A nation built on dreams and demand,
Opportunities abound, without a doubt,
For those who seek, and never give out.

So let us cherish this land we share,
With grateful hearts and love so rare,
For in the United States we find,
A beacon of hope, for all mankind.

TWENTY-FIVE

FOREVERMORE

In the land of dreams and golden shores,
Where freedom's flame forever soars,
A nation born of diverse lands,
United by courageous hands.
From sea to shining sea, behold,
A tapestry of stories untold.
Mountains majestically rise,
While valleys echo lullabies.
From the Rockies' snowy peaks,
To the Great Lakes where beauty speaks,
From the deserts' sunlit dunes,
To the forests' whispering tunes.
In the heartland's golden fields,
The American spirit never yields.
Where farmers toil day and night,
To feed a nation with all their might.

In the city that never sleeps,
Where dreams are born in endless heaps,
New York, a beacon of hope and light,
Where dreams take flight, reaching new heights.

From the bustling streets of L.A.,
To the southern charm of the deep bayou,
From the Midwest's humble grace,
To the Pacific's wild embrace.

United we stand, a nation so grand,
Hand in hand, we firmly withstand,
Every challenge that comes our way,
With justice and equality, we'll stay.

America, land of the brave and free,
Where dreams come true for you and me,
With boundless opportunities to explore,
The American dream forevermore.

TWENTY-SIX

MELTING POT HOME

In the land of stars and stripes, so grand,
Where dreams take flight and hope expands,
A nation built on freedom's call,
United we stand, together we fall.

From sea to shining sea we roam,
A tapestry of cultures, a melting pot home,
From the bustling streets of New York City,
To the majestic peaks of the Rockies so pretty.

Resilience and determination abide,
In the hearts of those who walk with pride,
Through trials and triumphs, we emerge strong,
For in unity we find where we belong.

From the southern states with vibrant soul,
To the Midwest plains where the prairies roll,
In the west, the golden coast so fair,
And the heartland's farms, a bountiful share.

From the jazz of New Orleans to Nashville's tune,
To the artistic energy of New York's commune,
Spires of innovation, reaching for the sky,
In the land of opportunity, dreams never die.

Oh, the hardworking farmers, day by day,
Providing abundance, nature's display,
And in the city that never sleeps, so bright,
A symbol of ambition, a beacon of light.

In this land of dreams, where opportunities bloom,
Where all can strive, where all can consume,
The United States of America, a land so grand,
A place where dreams come true, across this mighty land.

TWENTY-SEVEN

JOY AND CHEER

In the land of dreams and endless skies,
Where freedom's flame forever flies,
A nation formed with hopeful stride,
The United States, our cherished pride.
   From coast to coast, a tapestry unfolds,
A symphony of stories, legends untold.
In the East, where history resides,
Colonial towns with cobblestone sides.
   New York City, a beacon of might,
Skyscrapers reaching for the light.
Chicago's skyline, bold and grand,
A city that never fails to command.
   In the South, where the sun shines bright,
Swaying palms and warm, starry nights.

New Orleans dances to jazz's beat,
A melting pot of cultures, vibrant and sweet.
    Out West, where the mountains stand tall,
Nature's wonders enchant us all.
The Grand Canyon's awe-inspiring view,
A testament to Earth's wonders anew.
    Onward to the heartland, fields of gold,
Where farmers toil, stories unfold.
Their hard work feeds a nation's need,
Their spirit of resilience, a noble creed.
    From sea to shining sea, united we stand,
In diversity, we find strength in hand.
A land of dreams, where opportunities gleam,
The United States, the land of the free.
    So let us celebrate, with joy and cheer,
The land that we hold dear.
America, the beautiful and free,
A beacon of hope for all to see.

TWENTY-EIGHT

SACRED PLACE

In the land where dreams are born,
Where freedom's flame forever burns,
Stands a nation strong and true,
The United States, red, white, and blue.

From New York's streets so grand,
Where dreams are built with every hand,
To California's golden coast,
Where dreams take flight and never boast.

In the heartland's fields of gold,
Where farmers toil, their stories untold,
They sow the seeds of hope and pride,
With calloused hands, their dreams collide.

In the mountains tall and grand,
Where nature's beauty takes command,
We find a strength that runs so deep,
A unity that we shall keep.

    From the bustling cities to the quiet towns,
From the prairies wide to the ocean's sounds,
A tapestry of cultures, vibrant and true,
The melting pot of red, white, and blue.
    In every corner, a story unfolds,
Of dreams pursued and tales untold,
Opportunity's land, where hope is found,
With open arms, we all are bound.
    So let us celebrate this land so fair,
With hearts united, we shall dare,
To cherish freedom, love, and grace,
In the United States, our sacred place.

TWENTY-NINE

OPEN ARMS

In the land of dreams, where freedom soars,
United States, a nation adored.
From mountains grand to bustling streets,
A tapestry of cultures, where unity meets.
    From coast to coast, the prairies wide,
America's beauty, a boundless stride.
In golden fields and azure skies,
Hope and opportunity forever lies.
    Land of the brave, and home of the free,
A nation built on diversity.
Where dreams take flight, and hearts ignite,
In the land of promise, shining bright.
    From sea to shining sea, it's true,
America's spirit, strong and true.

With open arms, we welcome all,
A melting pot, where dreams stand tall.
    In the Southwest's embrace, the desert's fire,
A landscape ablaze, with dreams to inspire.
Rivers, lakes, and forests so grand,
Nature's wonders, across this land.
    From the red rock canyons to the rolling plains,
America's beauty forever remains.
With pride we stand, hand in hand,
United we are, a mighty band.
    In sunsets' hues and landscapes vast,
Strength and hope will forever last.
In this land of unity, we find our way,
America, land of dreams, we proudly say.

THIRTY

BOUNDLESS ADMIRATION

In the land of dreams, where freedom thrives,
A nation stands tall, where hope survives.
From sea to shining sea, a tapestry unfurled,
United in diversity, a beacon to the world.

From the snow-capped peaks of the Rocky Mountains,
To the rolling plains of the Midwest fountains,
The grandeur of nature, vast and untamed,
In America's embrace, every heart is claimed.

From the bustling streets of New York City,
To the peaceful shores of California's serenity,
A symphony of cultures, woven in harmony,
America's embrace, a testament to unity.

In the heartland, where amber waves abound,
The spirit of resilience, forever resounds.

From the deep South, where blues guitars cry,
To the wild West, where cowboys ride high.

    The spirit of freedom, a flame that burns bright,
Guiding us forward, through every fight.
From the heroes who fought for our liberty,
To the dreamers who shape our destiny.

    America, the land of promise and pride,
Where dreams take flight, and hopes collide.
In this land of opportunity, so vast and wide,
We stand together, with love as our guide.

    So let us celebrate this great nation,
With gratitude and boundless admiration.
For in America's embrace, we find our worth,
The land of the free, the home of our birth.

THIRTY-ONE

PROUDLY PROCLAIM

In the land of dreams and endless skies,
Where mountains peak and rivers rise,
There lies a land so vast and free,
A nation that's called America, you see.

From the shores of the East, where the sun first gleams,
To the golden coast, where the Pacific streams,
From the heartland's plains, where the wind whispers,
To the Rocky Mountains, where the spirit stirs.

In this tapestry of cultures and creed,
A symphony of voices, in harmony, they lead,
A melting pot of dreams, hopes, and desires,
An ode to the human spirit that never tires.

From sea to shining sea, united we stand,
In the face of adversity, hand in hand,

From the Founding Fathers' ink-stained quill,
To the heroes who fight for freedom still.
In the heartland, where fields stretch wide,
Where hard work and faith collide,
In the cities, where dreams take flight,
Where innovation sparks, igniting the night.
Across this land, beauty unfolds,
In national parks, where nature holds,
In sunsets that paint the sky with fire,
In the resilience that fuels our desire.
America, land of the brave and free,
A beacon of hope for all to see,
With flag unfurled, we proudly proclaim,
Our love for this land, our eternal flame.

THIRTY-TWO

UNITED, PROUD, AND FOREVER FREE.

In the land of dreams and golden shores,
Where liberty's flame forever soars,
Stands a nation unique and grand,
The United States, a wondrous land.
 From the Southwest's fiery embrace,
To the Northeast's autumn grace,
Each corner holds a different hue,
A tapestry of red, white, and blue.
 In the heartland, fields stretch wide,
Where hardworking souls reside,
Their labor sows the seeds of growth,
From coast to coast, a testament to both.
 Through the deep South's blues guitars,
To the Rockies' majestic stars,

A symphony of cultures thrives,
A melting pot where hope survives.
    The Founding Fathers, brave and wise,
Their vision, our eternal prize,
Heroes fight for freedom's call,
Their courage echoes through it all.
    In cities bustling with vibrant life,
Dreamers chase their hopes and strife,
In every street, a story unfolds,
In unity, our strength upholds.
    So let us raise our voices high,
With gratitude and love, we signify,
Our admiration for this great land,
For the United States, forever grand.
    America, the land of the free,
We sing your praise, eternally,
With hearts aflame, we'll always be,
United, proud, and forever free.

THIRTY-THREE

VITAL SEED

In the land where dreams are born,
A country of hope and freedom adorned,
United States, a nation bold and grand,
With landscapes vast and diverse, so grand.

From the towering peaks of the Rocky Range,
To the golden shores where oceans exchange,
Each state a tale of beauty and wonder,
A tapestry of landscapes, a sight to ponder.

In the heartland, fields of gold stretch wide,
Where amber waves of grain dance with pride,
And rivers carve a path through fertile ground,
A symphony of nature's harmony profound.

From the bustling cities that never sleep,
Where dreams are nurtured, ambitions leap,
A melting pot of cultures, colors, and creed,
A testament to unity, a vital seed.

Through strife and struggle, the nation stands,
Resilient, united, holding hands,
A beacon of hope, a guiding light,
A symbol of strength, shining so bright.
   America, land of countless dreams,
Where opportunity gleams and redeems,
With gratitude and pride, let us proclaim,
Our love for this nation, our sacred flame.
   For in the United States, we find,
A land where dreams take flight, unconfined,
A nation of courage, resilience, and grace,
America, let our hearts forever embrace.

THIRTY-FOUR

PEARL

In the land of the brave and the home of the free,
Where dreams take flight and hopes find glee,
Stands a nation, strong and true,
With skies of red, white, and blue.

From sea to shining sea, it stretches wide,
With landscapes diverse, a scenic pride,
Mountains that kiss the heavens' gate,
And plains that whisper tales of fate.

From New York's city lights so bright,
To California's golden shores, a stunning sight,
From the heartland's amber waves of grain,
To the southern charm that knows no pain.

A tapestry woven with colors bold,
A nation of stories yet untold,

A melting pot of cultures and creed,
Where unity thrives, and dreams succeed.

    Here, heroes rise, with courage unbowed,
Their valor and sacrifice, forever endowed,
From the founding fathers' noble quest,
To the pioneers who forged the West.

    In this land, where dreams take flight,
Where opportunity shines so bright,
With resilience and hope, we stand tall,
United, we'll conquer any fall.

    So let the stars and stripes unfurl,
As we celebrate this great nation's pearl,
A land of promise, a beacon of light,
The United States, shining ever so bright.

THIRTY-FIVE

THE GREATNESS OF THE UNITED STATES

In the land where the magnolias bloom,
And the sun paints the sky in golden hues,
Lies a place of charm and southern grace,
Where cultures blend in a vibrant embrace.
    From the bustling streets of New York City,
To the rolling hills of Tennessee,
From the shores of California's coast,
To the bayous where the alligators boast.
    This is the United States of America,
A tapestry of diversity and pride,
Where dreams are born and hopes collide,
A nation built on unity and resilience.
    From the mountains of Colorado's grandeur,
To the beaches of Florida's sunny shore,

From the deserts of Arizona's heat,
To the Midwest's fields of golden wheat.
  This is a land of endless opportunity,
Where courage and resilience pave the way,
Where dreams can manifest and come to play,
In the land of the brave and the home of the free.
  So let us celebrate this great nation,
With its beauty, freedom, and innovation,
From sea to shining sea, let us proclaim,
The greatness of the United States, our claim.

ABOUT THE AUTHOR

Walter the Educator is one of the pseudonyms for Walter Anderson. Formally educated in Chemistry, Business, and Education, he is an educator, an author, a diverse entrepreneur, and he is the son of a disabled war veteran. "Walter the Educator" shares his time between educating and creating. He holds interests and owns several creative projects that entertain, enlighten, enhance, and educate, hoping to inspire and motivate you.

Follow, find new works, and stay up to date
with Walter the Educator™
at WaltertheEducator.com

www.ingramcontent.com/pod-product-compliance
Lightning Source LLC
LaVergne TN
LVHW051958060526
838201LV00059B/3723